# Backyard Birds of the West

Genevieve Einstein
& Einstein Sisters

KidsWorld

# Quick Guide

These are some of the birds that you are most likely to see in your backyard if you live in Western North America. We've included their measurements in case you want to have fun with rulers!

**Rock Pigeon p. 8**
Length: 13 in (33 cm)

**Mourning Dove p. 10**
Length: 11 in (29 cm)

**Anna's Hummingbird p. 12**
Length: 4 in (10 cm)

**Rufous Hummingbird p. 14**
Length: 3 in (8 cm)

**Ring-billed Gull p. 16**
Length: 19 in (49 cm)

**Cooper's Hawk p. 18**
Length: 15 in (38 cm)

**Red-tailed Hawk p. 20**
Length: 20 in (51 cm)

**Great Horned Owl p. 22**
Length: 21 in (55 cm)

**Red-naped Sapsucker p. 24**
Length: 8 in (20 cm)

**Downy
Woodpecker p. 26**

Length: 6 in (16 cm)

**Northern
Flicker p. 28**

Length: 12 in (30 cm)

**American
Kestrel p. 30**

Length: 10 in (25 cm)

**Black-billed
Magpie p. 32**

Length: 21 in (55 cm)

**American
Crow p. 34**

Length: 18 in (47 cm)

**Barn Swallow p. 36**

Length: 7 in (18 cm)

**Black-capped
Chickadee p. 38**

Length: 5 in (13 cm)

**Red-breasted
Nuthatch p. 40**

Length: 4 in (10 cm)

**White-breasted
Nuthatch p. 42**

Length: 5 in (13 cm)

**House Wren p. 44**

Length: 5 in (13 cm)

**Ruby-crowned
Kinglet p. 46**

Length: 4 in (10 cm)

**Western
Bluebird p. 48**

Length: 7 in (18 cm)

**Hermit Thrush p. 50**
Length: 6 in (16 cm)

**American Robin p. 52**
Length: 10 in (25 cm)

**Varied Thrush p. 54**
Length: 9 in (23 cm)

**Gray Catbird p. 56**
Length: 9 in (23 cm)

**European Starling p. 58**
Length: 9 in (23 cm)

**Bohemian Waxwing p. 60**
Length: 7 in (18 cm)

**Cedar Waxwing p. 62**
Length: 6 in (16 cm)

**House Sparrow p. 64**
Length: 6 in (16 cm)

**Evening Grosbeak p. 66**
Length: 7 in (18 cm)

**Purple Finch p. 68**
Length: 6 in (16 cm)

**American Goldfinch p. 70**
Length: 5 in (13 cm)

**Chipping Sparrow p. 72**
Length: 5 in (13 cm)

**Dark-eyed
Junco p. 74**

Length: 6 in (16 cm)

**White-crowned
Sparrow p. 76**

Length: 6 in (16 cm)

**Spotted
Towhee p. 78**

Length: 8 in (20 cm)

**Western
Meadowlark p. 80**

Length: 8 in (20 cm)

**Red-winged
Blackbird p. 82**

Length: 8 in (20 cm)

**Brown-headed
Cowbird p. 84**

Length: 8 in (20 cm)

**Brewer's
Blackbird p. 86**

Length: 9 in (23 cm)

**American
Redstart p. 88**

Length: 5 in (13 cm)

**Yellow-rumped
Warbler p. 90**

Length: 5 in (13 cm)

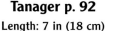

**Western
Tanager p. 92**

Length: 7 in (18 cm)

**Black-headed
Grosbeak p. 94**

Length: 7 in (18 cm)

# How to Use this Book

Each bird in this book has icons in the top right-hand corner. These icons quickly tell you the size of the bird, where to look for it, the food it eats and the kind of nest it builds.

## Size

Small is for birds that are shorter than the length of a school ruler (1 ft/30 cm).

Medium is for birds that are between one and two school rulers long (1-2 ft/30-60 cm).

Large is for birds that are bigger than two school rulers placed end to end (2 ft/60 cm).

## Where to Look

On or close to the ground

On or close to the water

In trees or shrubs

In the air

6

# Food

 Seeds, flowers or other plant parts

 Fruits or berries

 Insects or other creepy crawlies

 Fish or other water animals

 Land animals (like mice) or birds

# Nests

Simple nests are often on the ground. Birds don't put much effort into simple nests.

Cup-shaped nests are usually found in trees. These are often made with grass and twigs.

Some birds like to nest in tree cavities or nest boxes.

Some birds have nests that are unusual. They don't fit into the other categories.

# Rock Pigeon

Rock Pigeons are common in backyards and parks. Most birds are gray with shiny feathers on their necks, but Rock Pigeons can also be brown or white.

Rock Pigeons strut around on the ground, making cooing sounds. When attracting a mate, the male will also fan his tail and bow his head. He may even feed her "milk" from his crop, a pouch in his throat!

Baby pigeons are called squabs. When they first start to grow in, the feathers look a bit more like porcupine quills than feathers!

9

# Mourning Dove

The Mourning Dove eats seeds from the ground or from platform feeders. It also swallows small stones or grit to help it break down the tough seeds.

The Mourning Dove's wings whistle when it flies, especially during takeoff. A group of doves is called a flight of doves.

The Mourning Dove makes its nest in a tree or on the ground. It lays 2 eggs in a nest but can build more than one nest each year.

Both parents look after the young. Mourning Doves feed their nestlings crop milk, which they make from the seeds they eat. It is stored in a pouch in the throat called a crop.

# Anna's Hummingbird

The male Anna's Hummingbird has a shiny, red head. As he sings his song of buzzes and whistles, he turns his head side to side to show off his flashy feathers.

With its long tongue, Anna's Hummingbird drinks nectar from flowers and catches insects. It visits backyard sugar-water feeders, and it will snatch insects off spider webs if it has the chance!

The female Anna's Hummingbird raises her 2 chicks on her own. She feeds them nectar and insects.

Q: Why do hummingbirds hum?
A: Because they can't remember the words!

13

# Rufous Hummingbird

The Rufous Hummingbird visits backyard sugar-water feeders filled with a mixture of one part sugar to 4 parts water. It will fight off other hummingbirds, even those twice its size!

This hummingbird can beat its wings more than 50 times per second! It makes a humming sound as it flies. The pitch and loudness of the hum depends on how quickly its wings are beating.

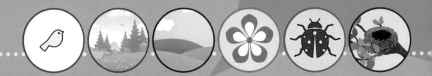

Unlike the male, the female is mostly greenish colored. She builds her apricot-sized nest out of soft plant materials glued together with spider webs. She decorates it with lichens or moss.

15

# Ring-billed Gull

The Ring-billed Gull is most likely to be seen in your yard in winter. At this time of year, its head has gray spots on it.

Ring-billed Gulls aren't picky about what they eat. They will eat all kinds of food left unattended by people, such as fruits, peanuts and even French fries!

Speckled **nestlings look** very different than adults.

In the spring and summer the adult Ring-billed Gull has only white on its head. In all seasons, it has the black ring at the tip of its bill that earned it its name.

17

# Cooper's Hawk

The Cooper's Hawk regularly visits backyards to find its lunch. It is interested in bird feeders, not for the seeds, but for the birds the seeds attract. It also eats squirrels and rabbits.

Q: What do you call a hawk with a lawn mower?
A: A mow-hawk!

The Cooper's Hawk also likes backyard birdbaths. After having a bath, it will usually move to a sunny perch to let the sun dry its feathers.

Young Cooper's Hawks are brown and white. All Cooper's Hawks have long tails and strong, sharp talons.

# Red-tailed Hawk

The Red-tailed Hawk sometimes visits large backyards to hunt for mice or squirrels. It often perches on a fence, using its sharp eyes to spot its lunch.

The red tail that this bird is named for is easily seen when it is flying. Young or very dark-colored Red-tailed Hawks don't have the red tail.

The Red-tailed Hawk may eat snakes, especially ones that aren't poisonous, like garter snakes. It will sometimes take its chances with a rattlesnake though!

21

# Great Horned Owl

The Great Horned Owl comes into backyards to catch feeder birds, mice or rabbits. This owl will even make a backyard nest box its summer home if the box is the right size.

The Great Horned Owl has very soft feathers. The soft feathers help keep the owl warm and also help it fly silently.

This owl finds its prey with its excellent hearing and its strong night vision. It can also turn its head to look directly behind itself!

Great Horned Owls usually use abandoned nests of other birds or cavities in trees. Sometimes they will even nest on buildings. A female can raise up to 4 owlets in her nest.

# Red-naped Sapsucker

If you see neat rows of small holes in a tree, there is a good chance a Red-naped Sapsucker has come to visit.

This bird likes to sip the sugary sap from the holes it makes in trees. It has hairs on its tongue that help it lick up the sap.

The Red-naped Sapsucker doesn't only eat sap. It also eats insects, especially those that get trapped in its sap wells! It also eats fruits, like berries.

Q: What did the tree say to the woodpecker?
A: Leaf me alone!

# Downy Woodpecker

Downy and Hairy woodpeckers look very similar. Size is the best way to tell them apart. The Downy Woodpecker is the size of a sparrow, and the Hairy Woodpecker is larger, the size of an American Robin.

Male Downy Woodpeckers have a red patch at the back of the head. Males spend their time on the smaller branches, and females spend more time on tree trunks and larger branches.

The Downy Woodpecker usually makes its nest where fungal infection has softened the wood. The male and female team up to carve out the nest hole, a task that can take up to 3 weeks.

The Downy Woodpecker's favorite backyard food is suet, but it will also eat peanuts, millet and black oil sunflower seeds. These are just treats, though. Insects make up most of its diet.

27

# Northern Flicker

The Northern Flicker's favorite food is ants, so it spends a lot of time on the ground. It also eats suet, sunflower seeds or peanuts at bird feeders.

Like most woodpeckers, the Northern Flicker has a really long tongue. The tongue is coated with sticky saliva that helps it catch the insects it likes to eat.

**The Northern Flicker usually carves out its own nest in a tree, but it can be attracted to a backyard nest box.**

**Nestlings are fed ant larvae that the parent carries in its crop.**

# American Kestrel

The American Kestrel is the smallest falcon in North America and also the most common. It may come to your yard to catch mice or insects, like grasshoppers and dragonflies.

The male has gray feathers in his wings. The female's wings are brown with dark bars. Both have a black mustache and black sideburns.

The American Kestrel makes its nest in abandoned woodpecker cavities or nest boxes. The female lays 4 to 5 eggs, and both parents take turns incubating the eggs and feeding the nestlings.

# Black-billed Magpie

The Black-billed Magpie has shiny blue-green feathers on its wings and tail.

This noisy bird regularly visits platform feeders or the ground beneath them. It will happily accept just about any backyard offering, including seeds, fruit, peanuts and suet.

Magpies gather together to find food, chase off hawks or steal food from larger predators like coyotes and foxes. A group of magpies is called a mischief of magpies.

Black-billed Magpie parents team up to build their domed nest. The male builds the outside, while the female makes a mud cup inside and lines it with grass.

33

# American Crow

American Crows are best known for making a caw sound. Really though, they have more than 20 different types of calls, which all have different meanings.

This bold, noisy bird is a regular visitor to backyards. It is really smart and can even tell different people apart.

You may see this bird stealing pet food from a dog dish, taking peanuts or fruit left out for other birds, flying off with your lunch or ripping open a garbage bag to eat the food scraps inside.

Q: Why did the crow stand on the telephone pole?
A: He wanted to make a long-distance caw.

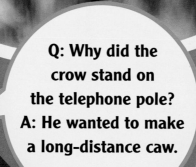

# Barn Swallow

The **Barn Swallow** eats mostly insects that it catches while flying. It is fond of eggshells, too. If you crush eggshells and place them outside, this swallow may come take them.

In flight, you can tell the Barn Swallow from other swallows because its tail is very forked. The outside feathers are much longer than the inside feathers.

The Barn Swallow used to nest in caves, but now it mostly makes its nest on buildings. It carries balls of mud in its bill and sticks them together, then lines the inside with dried grass and feathers.

Chicks leave the nest at about 3 weeks old. By about a month old, the young are able to catch their own food.

# Black-capped Chickadee

The **Black-capped Chickadee is a common backyard bird. It is easily attracted to feeders for sunflower seeds, suet and peanuts.**

When you hear this bird's **chick-a-dee-dee** call, you will know where its name comes from! If there are more **dee** notes at the end, it means the bird is trying to warn flock mates about some kind of danger.

The female often uses moss and fur to line her nest. Both parents catch insects to feed their young. The parents keep feeding their young for up to a month after they leave the nest.

The Black-capped Chickadee hides food for the cold winter months. It has a great memory and can remember where all its food is hidden!

# Red-breasted Nuthatch

The **Red-breasted Nuthatch** eats mostly insects in the summer and seeds in the fall and winter. It will happily take seeds, suet and peanuts from feeders year-round.

This bird makes a nasal **yank yank** sound. A breeding pair stays together all year long, and they often talk to each other in this way.

The Red-breasted Nuthatch often carves out its own nest cavity in the trunk of a tree. It smears the nest entrance with sticky tree resin. Sometimes this bird steals nest-lining material from other birds' nests!

# White-breasted Nuthatch

Nuthatches are nicknamed upside-down birds
because they often scurry down trees headfirst.
White-breasted Nuthatches do not have the black
eye band of the Red-breasted Nuthatch.

Nuthatches will hammer a nut with their bills to hatch the nut open. That's how they earned the name nuthatch.

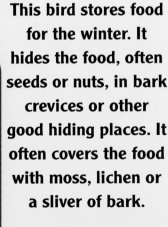

This bird stores food for the winter. It hides the food, often seeds or nuts, in bark crevices or other good hiding places. It often covers the food with moss, lichen or a sliver of bark.

**43**

# House Wren

The **House Wren** is a common backyard visitor. It searches for spiders and caterpillars in gardens and bushes.

The House Wren is a plain, brown bird with a beautiful, bubbly song. It usually holds its tail tilted up when perched or standing.

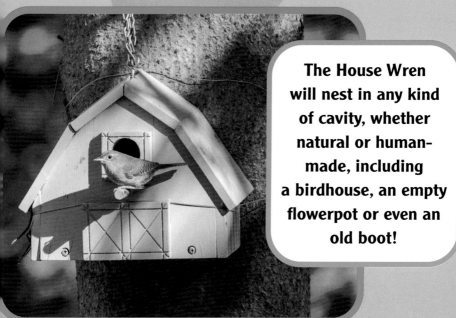

The House Wren will nest in any kind of cavity, whether natural or human-made, including a birdhouse, an empty flowerpot or even an old boot!

The House Wren lines its nest with feathers, grass and hair. Sometimes spider egg sacs are added to the nest as well!

45

# Ruby-crowned Kinglet

The Ruby-crowned Kinglet may visit your bird feeder for suet or sunflower seeds in the fall and winter. In the summer, it is more likely to visit your yard looking for insects and spiders.

Q: Why do boy birds look like their dad?
A: Like feather, like son!

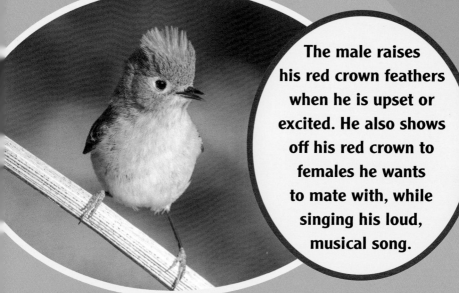

The male raises his red crown feathers when he is upset or excited. He also shows off his red crown to females he wants to mate with, while singing his loud, musical song.

The female Ruby-crowned Kinglet does not have the red crown. But both male and female birds have a white ring around their eyes and a white stripe on their wings. This restless bird flicks its wings many times per minute.

47

# Western Bluebird

In the summer, Western Bluebirds eat mostly insects. If you have seen them around, you can attract them to your yard by offering mealworms at your feeder.

The female Western Bluebird builds her nest inside a tree cavity or nest box. She lays up to 8 eggs and incubates them for about 2 weeks before they hatch. Both parents feed the chicks.

Western Bluebirds like to gather together in groups to bathe. They shake their wings and tail in the water, while flicking water with their bills.

49

# Hermit Thrush

The Hermit Thrush uses its foot to shake the grass and stir up insects that it catches and eats! It will eat fruit or mealworms if you leave them out in your yard.

The Hermit Thrush's song has an echoey sound. Although the songs of every Hermit Thrush sound the same to a human ear, each male puts his own unique spin on it.

Thrushes are robin-sized birds that all look similar, with brown backs and spotting on their breasts. The Hermit Thrush is different from other thrushes because its tail is reddish brown.

# American Robin

An **American Robin** may come to your bird feeder, but you are more likely to see it pulling earthworms out of your lawn.

Both the female and male have a red belly and gray upperparts, but the color of the female is paler overall. The robin's song sounds a bit like cheerily, cheer up, cheer up, cheerily, cheer up!

The female builds her nest and lays 3 to 5 eggs. Both parents feed the chicks. Nestlings beg for food by stretching their necks up high and opening their mouths wide.

Chicks leave the nest at about 2 weeks old, but the parents feed them until they are 5 weeks old.

# Varied Thrush

The Varied Thrush may visit your backyard in winter. It eats hulled sunflower seeds, suet, fruit and mealworms. A male may even defend a small territory around your platform feeder!

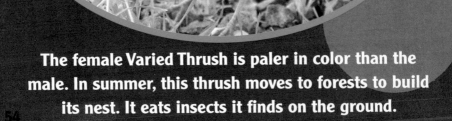

The female Varied Thrush is paler in color than the male. In summer, this thrush moves to forests to build its nest. It eats insects it finds on the ground.

The Varied Thrush may not look camouflaged in trees. On the forest floor, though, where orange leaves have fallen and conifer needles are scattered on the dark ground, this bird is harder to see.

# Gray Catbird

The Gray Catbird doesn't often visit feeders, but it will take fruit or mealworms if they are offered. This gray bird has a black cap and a cinnamon brown patch under its tail.

This bird makes a mewing call that sounds a lot like a cat. Its song can include imitations of other birds as well as mechanical sounds. And it can sing with 2 voices at once!

The Gray Catbird comes to western North America in the spring and summer to nest. Young catbirds leave the nest at about 10 days old. Chicks are fed insects and spiders.

57

# European Starling

In the fall, the European Starling grows new feathers that have whitish tips. By the time spring rolls around, the tips have worn off, leaving only the glossy black part of the feathers.

The European Starling was introduced to North America about 130 years ago. Now it is common over most of North America. It will visit backyard feeders and isn't fussy about what it eats.

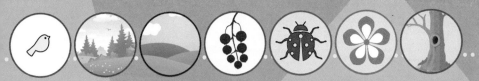

European Starlings like to gather in groups and are good imitators. They often imitate the songs of other bird species. Like parrots, they can even mimic human speech.

European Starlings are cavity nesters, but they don't limit themselves to tree cavities. They can also nest in buildings, holes in cliffs and sometimes even in burrows!

# Bohemian Waxwing

Bohemian Waxwings eat a lot of fruit. The waxy red or yellow feather tips on their wings and tail come from chemicals in the fruits they eat. Older birds have larger waxy patches on their feathers.

Large groups of waxwings gather to search for food together. They often take over a single tree to dine on the fruits, talking to each other with their high-pitched calls.

A Bohemian Waxwing chooses its mating partner in the winter. As part of courtship, the pair will pass a berry back and forth.

Q: What's a bird's favorite game?
A: Beak-a-boo!

61

# Cedar Waxwing

Most birds will remove seeds from
fruits, but not the Cedar Waxwing. It
just swallows them and poops them out.

Not all birds need to drink water, but the Cedar Waxwing needs to drink water or eat snow to balance out the high sugar content of the fruit it eats. It also really likes to take baths!

Cedar Waxwings nest in loose groups of 12 or so birds. To save time, sometimes females will steal nesting material from the nests of other bird species.

63

# House Sparrow

House Sparrows were introduced to North America from Europe more than 150 years ago. Now they are common in cities and towns across Canada and the U.S. They often visit backyard feeders.

The male House Sparrow has a black throat and bib, and the back of his neck is brown. The female is brown and beige overall, with a beige eye stripe.

House Sparrows often look for food in groups. Males with larger black bib patches are older and have a higher rank within the group.

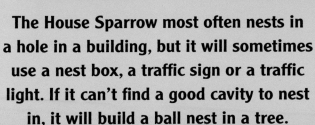

The House Sparrow most often nests in a hole in a building, but it will sometimes use a nest box, a traffic sign or a traffic light. If it can't find a good cavity to nest in, it will build a ball nest in a tree.

65

# Evening Grosbeak

You are most likely to see an Evening Grosbeak at your feeder in the winter. It likes to eat sunflower seeds, which it easily cracks open with its sturdy bill.

Q: What is a bird's favorite movie?
A: Lord of the Wings!

Even though it is considered a songbird, the Evening Grosbeak doesn't really have a song. It does make a variety of calls, including trills and chatters.

The female Evening Grosbeak has yellow only on her shoulders, but she has the same strong, thick bill as the male.

# Purple Finch

A famous birder once described the male Purple Finch as a "sparrow dipped in raspberry juice." This bird likes sunflower seeds and cracks them open easily with its strong bill.

The female Purple Finch is mostly brown with a white eyebrow and chin. As well as seeds, the Purple Finch also eats fruit and insects.

The male sings many different songs. He may even include portions of songs from other birds, such as the American Goldfinch or Eastern Towhee in his own song.

69

# American Goldfinch

The American Goldfinch is a seed eater that often visits bird feeders. If you have sunflowers or thistles in your yard, you might see it perched right on the flower eating the fresh seeds.

In the summer, the male is bright yellow with a black cap. The female has yellow on her belly, but her back and head are more olive green colored. Both have black wings and tail.

The call of the American Goldfinch sounds like po-ta-to-chip. In the spring, the male sings a song of twitters and warbles to help him attract a mate.

In the winter, even the American Goldfinch male has a much drabber, grayer plumage. Birds in the north migrate south in the winter to avoid freezing weather.

71

# Chipping Sparrow

You can tell the Chipping Sparrow apart from other sparrows that visit your yard by its reddish brown cap. It regularly eats seeds from feeders or on the ground.

The Chipping Sparrow generally eats insects only in the summer when it is raising its young.

Nestlings are generally fed seeds and insects by both parents. The young leave the nest when they are around 10 days old. Their parents continue to feed them for 3 weeks after they leave the nest.

The Chipping Sparrow is named for its chip contact call. It uses this call to stay in contact with family and flock mates.

# Dark-eyed Junco

The Dark-eyed Junco is a common visitor to backyards and bird feeders, especially in the winter. It eats a variety of seeds, nuts and grains.

The Dark-eyed Junco comes in different color forms. All forms have white outer tail feathers that are visible in flight. The form most common in the West has a brown back.

The female usually builds her nest on the ground. Some nests are well built, and others are flimsy. She lays 3 to 6 brown-spotted eggs in her nest.

After she incubates the eggs for about 12 days, the nestlings hatch. Both parents feed nestlings predigested food or insects.

75

# White-crowned Sparrow

You may see the White-crowned Sparrow at your feeder, but you are more likely to see it on the ground. It often cleans up seeds that fall below the feeder. Look for its black-and-white striped head.

Young White-crowned Sparrows have brown and gray stripes on their head.

This bird has a whistling song, and it makes a short penk call.

White-crowned Sparrows defend their territories, which may include your bird feeder! What starts as a singing duel can end with birds flying at each other and jabbing each other with their feet.

77

# Spotted Towhee

The Spotted Towhee likes sunflower seeds and grains. It prefers to feed on the ground, so if you see this bird your yard, consider sprinkling some seeds on the ground for them.

The Spotted Towhee doesn't only eat seeds. In the summer it also eats a lot of bugs, like caterpillars, beetles, grasshoppers and spiders.

Q: What do you give a sick bird? A: Tweetment!

Females and males look similar, but the male is more richly colored and the female is more drab. Juveniles are brownish gray with pale orange markings.

# Western Meadowlark

If you live near grasslands, the Western Meadowlark may show up in your backyard. It eats seeds or grains that are on the ground, or it looks for insects.

The Western Meadowlark is easy to recognize. It has a black, V-shaped band on its yellow breast. The male sings a whistling, gurgling song.

The Western Meadowlark eats insects in the summer when they are available. It searches for insects by poking its bill into the ground to make a hole.

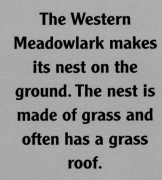

The Western Meadowlark makes its nest on the ground. The nest is made of grass and often has a grass roof.

81

# Red-winged Blackbird

The male Red-winged Blackbird is easy to recognize with his bright red-and-yellow wing patch and his loud conk-la-ree call.

The female is brown and buff with a pale eyebrow. The Red-winged Blackbird eats mostly insects in the summer and mostly plant food (often seeds or grains) in the winter.

When they hatch, nestlings cannot see and have very little down to keep them warm. Two weeks after hatching, when they leave the nest, they can see well and have a full set of flight feathers.

# Brown-headed Cowbird

The male Brown-headed Cowbird has a brown head and a black body. The female is pale brown. This bird eats seeds, grain or peanuts from backyard bird feeders.

This bird earned the name cowbird and the nickname Buffalo Bird because it likes to be near livestock. It eats the insects that these animals flush up as they move.

The Brown-headed Cowbird female lays her eggs in the nests of other birds and lets those birds raise her chicks. She can lay up to 40 eggs per season in many, many nests!

Brown-headed Cowbird chicks hatch early and grow quickly. They often grow to be larger than their adoptive parents who are raising them!

85

# Brewer's Blackbird

**Brewer's Blackbird is nicknamed Glossy Blackbird because of the male's shiny plumage. This bird might show up on your lawn to eat seeds or grain under your feeder, or to feed on grasshoppers or beetles it finds in your yard.**

The female is a drab brown color with only a slightly glossy back that is not very noticeable.

Brewer's Blackbirds build their nests in trees, shrubs or on the ground, but always in groups. A breeding colony may have few pairs, or it may have 100 pairs nesting close together!

# American Redstart

This little songbird won't come to your feeder, but it might come to your yard to dine on insects or small berries on your trees.

Q: What did one egg say to the other egg?
A: Let's get cracking!

American Redstarts fan their tails to stir up resting insects. Then they eat the insects that try to fly away. The female is yellow and gray instead of orange and black.

The female chooses the nest site and builds her nest. She looks for a location that will be protected from the rain and the sun and that will be harder for predators to notice.

# Yellow-rumped Warbler

The Yellow-rumped Warbler may visit your yard from April to October. It spends the winter in Mexico or Honduras.

The male is darker overall than the female with more black on his chest. Both males and females have yellow on their rump, throat and sides.

The female is less colorful overall, with grayer streaking on her chest.

The Yellow-rumped Warbler may also come to your yard for a bath. If you have a birdbath, it is important that the water is changed every few days to keep the water clean.

# Western Tanager

The **Western Tanager** will accept offerings of freshly cut oranges, as well as other fresh or dried fruits. The flashy male gets his red color from eating insects that eat certain plants.

The female has a bit of red only on her face and is less brightly colored overall. She builds her nest and incubates the eggs on her own. When the nestlings hatch, the male helps her feed them.

This bird may gather in loose flocks in migration. A group of tanagers is called a season of tanagers.

93

# Black-headed Grosbeak

The Black-headed Grosbeak is one of only a few birds that can eat the poisonous monarch butterfly.

Q: Why do birds fly south for the winter?
A: Because it's too far to walk!

Pairs start nesting in April or May. They eat a lot of insects in the summer, but they also visit feeders for sunflower seeds.

Females and young birds look similar. The adult males migrate south to Mexico first, so it is the females that feed the young later in the summer.

The Publisher: KidsWorld Books

**Library and Archives Canada Cataloguing in Publication**

Title: Backyard birds of the West / Genevieve Einstein & Einstein Sisters.

Names: Einstein, Genevieve, 1977– author. | Einstein Sisters, author.

Identifiers: Canadiana (print) 20210222794 | Canadiana (ebook) 20210222808 | ISBN 9781988183282 (softcover) | ISBN 9781988183299 (PDF)

Subjects: LCSH: Birds—Canada, Western—Identification—Juvenile literature. | LCSH: Birds—West (U.S.)—Identification—Juvenile literature. | LCSH: Birds—Canada, Western—Juvenile literature. | LCSH: Birds—West (U.S.)—Juvenile literature. | LCGFT: Field guides.

Classification: LCC QL685.W47 E46 2021 | DDC j598.09712—dc23

*Photo credits*

*Front cover:* GettyImages: michaelmill. *Back cover:* GettyImages: hannurama, mtruchon, SC Shank.

*Bird Illustrations:* Gary Ross, Ted Nordhagen, Ewa Pluciennik, Horst Krause

*Image Credits:* From GettyImages: Alberthep 29; alukich 95b; Angelika Koehne 93b; ArmanWerthPhotography 90b; bazilfoto 59a; BethWolff43 75b; blightylad-infocus 26a; Bob Balestri 47b; BrianEKushner 58a; BrianLasenby 44a, 56b; ca2hill 22b, 82a; Carol Hamilton 45b 89b; Chiyacat 70b; ChrisBoswell 39a; Christiane Godin 16a; Comstock Images 23a; DavidByronKeener 82b; davidehaas383 71b; Deborah Roy 20; Devonyu 31 40b; drakuliren 36b; ebettini 84b; Edward Palm 37a, 70a; EJ_Rodriquez 58b; emarys 33a; epantha 50b; Flatcoater 74a; hannurama 60a; Heather Burditt 37b, 57; Hurricane 68; impr2003 76a; Janet Griffin-Scott 10b, 11b; Jason Erickson 77a; Jeff Edwards 44b; JeffGoulden 38b; JenDeVos 39b; kahj19 81a; Ken Griffiths 28b; kojihirano 32a; LorraineHudgins 62; M. Leonard Photography 25; Mason Maron 12b, 52a; MichaelStubblefield 51; MikeLane45 61; mirceax 47a, 73b; Motionshooter 19b; mtruchon 53a; OldFulica 12a; Oren Ravid 65b; PamSchodt 85b; passion4nature 30b, 83; PaulFleet 64b; PaulReevesPhotography 21b, 40a, 50a, 67ab, 76b; Peter Milota Jr 18; phototrip 60b; R Lolli Morrow 27a; Ralph Navarro 45a; RCKeller 71a; Rebecca Turner 38a, 93a; RLSPHOTO 42; RONSAN4D 95a; SC Shank 30a; Sen Yang 13, 14a; shellhawker 65a; Sloot 64a; spates 52b; SteveByland 56a, 88; Supercaliphotolistic 21a; Tempau 16b; TenleyThompson 92; Thomas Anderson 26b; Thorsten Spoerlein 59b; Warren_Price 53b, 63a; William Krumpelman 36a; yhelfman 75a; zhuclear 11a. From Flickr: Alan Schmeirer 69a, 94; Allan Hack 8; Becky Matsubara 32b, 34b, 78a, 79; Care_SMC 46; Charles Gates 54b; CheepShot 17b; Courtney Celley USFWS 10a, 27b; cuatrok77 34a, 35; Darren Kirby 86; David A Mitchell 41, 43b; David Slater 9a; Dennis Murphy 73a; devra 91b; Don Debold 48b; Eric Gropp 55; F D Richards 72b; FancyLady 63b; fishhawk 66; Francesco Veronesi 78b; InAweofGod'sCreation 24ab; Jacob W Frank 15; Jeremy Meyer 89a; Jerry Kirkhart 80a, 91a; John Liu 9b; Ken Bosma 23b; Krista Lundgren USFWS 85a; malibuskiboats 72a; Mike's Birds 19a, 49ab; nature80020 22a, 33b, 80b, 84a; Nigel 48a; Paul VanDerWerf 90a; Richard Griffin 69b; Richard Masoner 87a; sam may 54a; Shawn McCready 28a; Sjensen 77b; Thomas Quine 74b; Tom Koerner USFWS 87b; USFS Pacific Northwest 81b; USFWS Mountain 17a; William Garrett 14b; Yanech Gary 43a.

Icons: GettyImages: Alexander_Kizilov; ChoochartSansong, rashadashurov, MerggyR, agrino, lioputra, FORGEM, Stevy, Intpro, MaksimYremenko, Oceloti, Thomas Lydell, Sudowoodo.

We acknowledge the financial support of the Government of Canada.
Nous reconnaissons l'appui financier du gouvernement du Canada.

Funded by the Government of Canada
Financé par le gouvernement du Canada | Canadä

PC: 38-1